LSD
AND THE
DIVINE SCIENTIST

"In this collection, Albert Hofmann shares his scientific, philosophical, psychological, and religious insights in an especially candid and engaging manner. His warmth, candor, and strength of character shine in them all, and through them we gain a satisfying glimpse into a great scientist's life."

RICK STRASSMAN, M.D., AUTHOR OF
DMT: THE SPIRIT MOLECULE AND COAUTHOR OF
INNER PATHS TO OUTER SPACE

"An extraordinary quality of Albert Hofmann was his ability to integrate the spiritual and the material under the conceptual umbrella of the natural. As his last work, *LSD and the Divine Scientist* presents Hofmann's most mature thought, the flowering of the scientist as philosopher in a natural integration of science and philosophy. May these final essays be the key to achieving Hofmann's ultimate vision."

NEAL GOLDSMITH, AUTHOR OF
PSYCHEDELIC HEALING

"This book is a jewel box of the many facets of Hofmann's mind: as a pure scientist and still in love with the discipline, as a science mystic making a clear-eyed case for the need for the two to blend, as a nature mystic drawing out the implications envisioned and verified in altered states, and as a philosopher of science grappling with the hard questions of death and the fabric of reality itself. Each essay was a rare mix—profound *and* readable. The essay by Alex Grey about the symbolism in his cover portrait of Hofmann is the perfect conclusion."

James Fadiman, author of
The Psychedelic Explorer's Guide

"Albert Hofmann's creation of LSD and identification of psilocybin had and continues to have a deep and lasting impact on psychiatry, and even more so on Western culture and spirituality. This invaluable collection reveals how, in his personal life, Albert leaned more toward spirituality than psychotherapy. In *LSD and the Divine Scientist,* Albert Hofmann reveals through his life and his work that we can be involved in spiritual practices while embracing the rigor of scientific methodology and a healthy, productive lifestyle. This inspiring book reveals how fortunate the world is that Albert was the father of LSD and, in profound ways, the spiritual father of millions of responsible psychedelic users."

Rick Doblin, president and founder of the
Multidisciplinary Association for Psychedelic
Studies (MAPS) and author of *Manifesting Minds*

LSD
AND THE
DIVINE SCIENTIST

The Final Thoughts and Reflections
of Albert Hofmann

ALBERT HOFMANN
Translated by Annabel Moynihan

Park Street Press
Rochester, Vermont • Toronto, Canada

Park Street Press
One Park Street
Rochester, Vermont 05767
www.ParkStPress.com

Park Street Press is a division of Inner Traditions International

Originally published in German under the title *Tun und Lassen: Essays, Gedaken und Gedichte* by Nachtschatten Verlag AG
First U.S. edition published in 2013 by Park Street Press

Library of Congress Cataloging-in-Publication Data
Hofmann, Albert, 1906–2008.
[Tun und Lassen. English]
LSD and the divine scientist : the final thoughts and reflections of Albert Hofmann / Albert Hofmann ; foreword by Christian Rätsch ; translated by Annabel Moynihan. — First U.S. edition.
pages cm
"Originally published in German under the title Tun und Lassen: Essays, Gedaken und Gedichte."
Includes bibliographical references and index.
ISBN 978-1-62055-009-0 (pbk.) — ISBN 978-1-62055-140-0 (e-book)
1. LSD (Drug) 2. Hallucinogenic drugs—Social aspects. I. Title.
BF209.L9H63313 2013
154.4—dc23

2012043658

Printed and bound in the United States

10 9 8 7 6 5 4 3 2

Text design and layout by Brian Boynton
This book was typeset in Garamond Premier Pro with Copperplate and Agenda as display typefaces

Contents

The Spiral

In the endless nothingness
is an imaginary point,
the start of every spiral:
the spiral
of the galaxies,
the ammonite,
the double helix.

In spirit, the perfect spiral leads
to eternity.
In time and space,
all spirals are bound,
cut off,
yet they too extend toward
infinity.

Translator's Note

Two years before he passed away, Albert Hofmann celebrated his hundredth birthday in good health, spirits, and mind. More than three thousand people from around the world came to Basel, Switzerland, to honor him, including this translator. The events and the attendees represented a colorful array of the topics and themes that spiral around Albert Hofmann. The book at hand touches on many of these themes. At one hundred years of age, Hofmann was sprightly and bright, and his delighted attitude about life shines through in these essays.

The spiritual core of these essays and poems is an understanding of how subjective reality is inextricably intertwined with the objective world. Reality is created in the mind, which, with its neurological pathways, sensory organs, and other physiological elements, is itself part of the measurable, objective world. For Albert Hofmann the intersection between these two worlds is a place of wonderful mystery.

I have long been drawn to Hofmann's enchanted view of the world we live in. As a graduate student I had the opportunity to translate one of Hofmann's essays for a class. After graduating my first translations were of Christian Rätsch's revisions to Hofmann and Richard Schultes's book *Plants of the Gods*. Since then I have worked on additional books and essays about psychoactive plants, including *The Encyclopedia of Psychoactive Plants* by Christian Rätsch.

Translators are faced with many conundrums, and one of the most common in German is what to do with the word Geist, which means both "mind" and "spirit." Like others, I have chosen one or the other based on the context and my own personal logic. Sometimes the choice is clear, while at other times the choice of "spirit" over "mind" seems as if it would add a superfluous intangibility. "Mind" intuitively sounds more concrete, but of course it isn't—thus the choice is never perfect, but such is the nature of translation. I would like to thank Deborah Loring, David E. Nichols, and Michael Moynihan for their assistance with this translation, and Jonathan Ott's translation of Albert Hofmann's *LSD: My Problem Child,* an invaluable source for correctly translating the chemical processes involved in the production of lysergic acid diethylamide.

I hope that this translation is faithful both to the objective words of the text and the subjective spirit of their author.

ANNABEL MOYNIHAN

Action and Allowing

Action and allowing are bound to time,
their past and their future are
in the material world.

Love and joy are the infinite nothingness
out of which the world is created
and re-created again.

Through love and joy we are,
each of us,
bound to the Creator

and we become creative
in the material world
with our action and our allowing.

FOREWORD

"We are all here to experience happiness," Albert Hofmann liked to say when asked about the meaning of life. His answer sounds so simple and easy, like the Socratic imperative to "Know thyself!"

One of Albert Hofmann's strengths and attributes was his ability to formulate philosophical ideas with a certain playful ease, using short and simple, yet profound, words. *Tun und Lassen* (Action and Allowing), the German title of this collection of aphorisms, poems, and articles, is itself such a disarmingly simple and yet complex phrase. A poem that Albert Hofmann wrote in 1993 and sent to his closest friends begins with those words.

He was a mystical friend of nature who kept an eye open for what Ernst Haeckel called the "art forms of nature." As an observer not a spectator, he recognized the universal creative principle of nature.

With enthusiasm he would examine the variations of the spirals in the ammonites and in the bewitchingly beautiful nautilus shells he collected. Most of all he loved the Japanese sea snail (*Pleurotomaria hirasei*) and the red triton snail (*Cymatium hepaticum*).

He saw the spiral as a spiritual principle, as the binding element of matter and spirit, of nature and culture, and as a self-creating power: a guide to the infinite. He expresses these ideas in his poem "The Spiral."

As a natural scientist, Albert Hofmann viewed nature itself as the primary material of the mystery. He was fond of saying, "If a chemist does not become a mystic, then he is not a chemist either." He understood the natural sciences as a mystical experience. He placed a great value on "chance."

For him, the final mystery of death—the "great transition"—was a natural phenomenon. He could only speculate about the answer to the question "What is death?" In the meantime, he has passed on. For him, everything is clear. For us, it still remains a mystery.

Christian Rätsch
Hamburg, Germany

CHRISTIAN RÄTSCH, PH.D., is a world-renowned anthropologist and ethnopharmacologist who specializes in the shamanic uses of plants for spiritual as well as medicinal purposes. His varied fieldwork has included research in Mexico, Thailand, Bali, Korea, the Peruvian and Columbian Amazon, as well as a long-term study on shamanism in Nepal. Before becoming a full-time author and lecturer, Rätsch worked as a professor of anthropology at the University of Bremen. He is the author of numerous articles and more than forty books including *Plants of Love, Gateway to Inner Space, Marijuana Medicine, The Dictionary of Sacred and Magical Plants,* and *The Encyclopedia of Psychoactive Plants* and *The Encyclopedia of Aphrodisiacs.*

The art forms of nature are the dreams of matter.

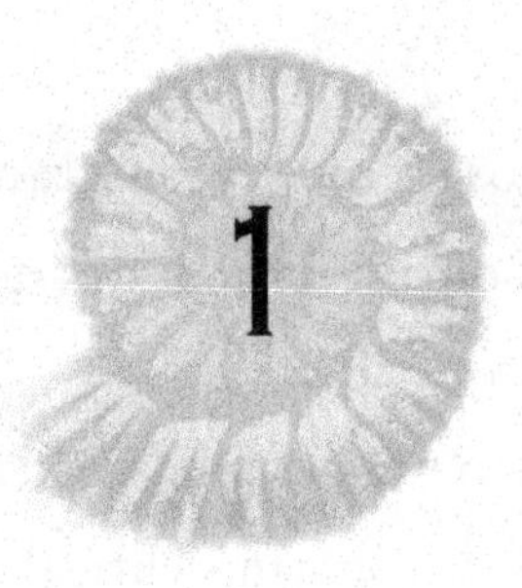

1 Planning and Chance in Pharmaceutical and Chemical Research

This is a slightly abbreviated manuscript of a lecture presented on January 17, 1979, at the Society for the Study of Nature (Naturforschende Gesellschaft) in Basel, Switzerland. It was first published in the journal *SwissPharma,* vol. I, no. 9 (1979).

This essay was written for an intended audience that extends beyond my colleagues in pharmaceutical research, for whom the following facts go without saying. In it I will talk about the various ways new medicines are developed and about the opportunities available to those engaged in targeted research, including the planning involved, and about the role that chance plays in all of this.

With regard to the complicated chemical formula of a medication, the pharmaceutical chemist is often asked questions like the following: How was a particular substance, with a certain chemical structure, first discovered? How could it have been known that such a compound would have a specific therapeutic effect? The answer to these questions is that it was not known in advance. Researchers could only establish that the substance was an effective medicine for patients after they had gathered pharmacological evidence in animal research, which then made further trials on humans seem reasonable.

We don't know *why* a particular chemical structure is endowed with a certain pharmacological effect. An impressive amount of knowledge about the mechanisms of actions of known pharmaceuticals has been collected throughout the world. This is the result of an enormous effort in scientific research using increasingly refined techniques and expensive methods and instruments. Because of this research scientists have been able to determine the biochemical and electrophysiological pathways through which the chemicals operate and ascertain to which structures of the organism they attach themselves. Today we know *how* many medications work, but we still know nothing about *why* they work in such a way. The relationship between a substance's chemical structure and its pharmacological effects does not follow any rules. *Ultimately, all knowledge regarding the relationships between chemical structures and pharmacological effects is based on empirical evidence.*

But whence originates this empirical knowledge? How is it gained? It has four sources.

1. It comes from ancient sources of folk medicine and is the result of researching active ingredients in medicinal plants.
2. It is the result of modern biological research and the analysis of physiologically active pharmaceutical ingredients.
3. It is the result of pharmacological screening of large numbers of synthetic compounds and natural ingredients.
4. It comes from observation of the medicine's effects on the patient.

MEDICINAL PLANT RESEARCH

We shall begin with a few remarks on the first empirical source mentioned above. Up until about a hundred years ago nearly all medications were of plant origin; a few were prepared from minerals or animals. The famous physicians of antiquity, the Middle Ages, and the Renaissance were all knowledgeable about herbal plant medicine. Asclepius, the Greek god of the physicians, was initiated into plant medicine by the centaur Chiron in herb-filled meadows on Mount Pelion. Until the rise of pharmaceutical chemistry this treasury of medicinal plant lore was contained in extensive encyclopedias and pharmacopoeias. These were based on the herbals of antiquity, especially those by Dioscorides in the first century CE and Galen in the second century CE.

To whom are we indebted for the knowledge contained in these books about the effects of certain plants? The writers are unknown, nameless explorers. Perhaps this knowledge was discovered during the search for nutritional food in the plant world. It is often assumed that the original humans were able to recognize the medicinal powers in plants and knew how to use them due to their good instincts. In the early years of scientific and chemical pharmaceutical research, scientists could reach back to this source of wisdom, and today research projects continue to be based on this knowledge.

Knowledge about the narcotic and analgesic effects of opium poppy resin inspired the pharmacist Friedrich Sertürner to look for its active ingredient—a search that led to the discovery of morphine. This was the first time an alkaloid had been isolated. Countless alkaloids and nitrogen-free substances were subsequently isolated in pure form by the targeted search for the active principle in other medicinal plants. Many valuable drugs—be they genuine natural substances or chemical modifications of a natural substance—have been discovered in this manner and thus found their way into the medicine cabinet.

Just a few examples of drugs discovered with this method include morphine, atropine, quinine, digitoxin, ergotamine, reserpine, and so forth. Today this previously fertile source of new and valuable pharmaceuticals has been largely exhausted, because the most important of the long-treasured ancient medicinal plants have already been investigated.

Naturally, pharmacologists are still needed for the development of medicine. They take the active ingredients, which chemists have isolated or chemically altered, and test their effects and toxicology in animal research. In clinical research physicians test the active compounds on humans and establish the correct medical indication and dosage.

As the title of this essay suggests, I wish to limit myself as a chemist to the description of the *chemistry* aspect of pharmaceutical research. Biology, physiology, pharmacology, and medicine play corresponding roles in this research.

PHYSIOLOGICALLY ACTIVE INGREDIENTS

We will now proceed to the second source of medicine mentioned above: physiologically active ingredients. Here too the chemist must begin his or her work with a substance that has a known effect. In contrast to the situation with traditional medicinal plants, this knowledge does not derive from ancient sources but is the result of modern biological research.

In this case physiologists and pharmacologists base their chemical research on investigations into the functions of the various organ systems, as well as of microorganisms. Such investigations led, for example, to the isolation of insulin, adrenaline, sexual hormones, adrenocortical hormone, pituitary hormones, and so on. Important medicines have been developed as a result. The discovery and isolation of vitamins, as well as of penicillin, also came about via this branch of pharmaceutical research.

Using advanced methods and a refined array of instruments, researchers today are able to penetrate further into the biochemical state of the various organs. As soon as discrete functions can be localized then it is the rewarding task of the chemist to isolate the physiologically active elements, elucidate their structure, and, when possible, synthesize and chemically modify them.

This is one of the most important and promising fields of research now available to scientists for discovering new active ingredients that might be used as medicine. One important example has been the discovery of the analgesic action of endorphin-type peptides. The medications derived from this area of research are also valued, because their physiologically active ingredients are of a type that is recognized by the body itself.

Common to both of the above methods for discovering new medicines is the fact that the research is based on a known effect for which the corresponding chemical structure is determined through the isolation and structural clarification of its active ingredients.

PHARMACOLOGICAL SCREENING

The other principal way in which researchers proceed by starting with a known structure and then looking for the corresponding effect is in synthetic chemistry research. This presents a third method for attaining an understanding of the relationship between the structure and the effect.

A chemist engaged in synthetic research has two procedural possibilities, depending on the screening program in which he or she is working. If the research is being done within the framework of a general screening program that encompasses all possible pharmacological, biological, and microbiological tests, then there are no limits to a chemist's imagination about the synthesis of new combinations. The chemist will want to synthesize as many new chemically original or at least pharmacologically untested substances as possible. If and how these substances will work is not known. It is pure chance whether anything viable will remain after the pharmacological screening. Instant successes in the form of a new type of structure and action are rare but not impossible; the best examples are the discovery of 1,4-benzodiazepines and the formulation of Librium (chlordiazepoxide).

Much more common than such general-screening programs, however, are targeted programs that only test a substance for specific activity. For example, a substance is tested for its psychotropic effects, or its effect on circulation, on insulin, and so forth. The chemists involved in such targeted research are no longer able to freely synthesize. For the project at hand they must limit themselves to synthesizing substances that can be expected to have a relevant targeted effect: structures with characteristics and particular effects that are known either from personal experience or through the research literature. They then attempt to make molecular modifications—which have not yet been tested—to these structures. Inventive new structure-effect models are less likely to be discovered in this line of research, but it does lead to the improvement of existing medicines.

PATIENT OBSERVATION

The fourth source of empirical knowledge about a drug's properties is the observations made by physicians of their patients.

After researchers have discovered a pharmacological action in a new substance through one of the three aforementioned methods, they must clear yet another hurdle before it can be accepted as medicine: they must test it as such on the patient. However, because the pharmacological effects of a particular substance will usually vary considerably from species to species, quite often a different sort of effect manifests when it is tested on humans from what had been expected based on the animal research.

In addition to the unpredictability of a substance's effects on a biological subject based on its chemical structure, there is the unpredictability of its effects on humans based on its activity in animal research.

As a result, only a very small percentage of experimental preparations that end up in clinical tests prove to have a therapeutic value for humans.

The regulatory requirements that experimental preparations are subject to before they are allowed in clinical tests have increased tremendously in recent years. This has come about due to past cases of drug toxicity, the worst example of which was the thalidomide affair. These cases did not usually come about due to poor procedural processes but rather the fact that only a conditional inferability exists between pharmacological effects on animals, including toxicity, and humans. The extensive data on a preparation—about its metabolism and toxicity and its effects on *various* animal species—that must be submitted before it is approved for clinical testing has driven up the costs of preclinical research so as to make it nearly impractical. This causes a reduction in the number of the active substances that can be tested on humans and a consequent reduction of the chances for discovering new medications. Yet even the most comprehensive testing on animals cannot exclude the possibility of negative side effects in humans.

The following are a few examples of drugs with unexpected positive effects on patients.

- 1-isonicotinoyl-2-isopropyl hydrazine (Iproniazid) is a substance that, on account of its bacteriostatic effect, was initially used as a tuberculosis drug. It was revealed in practice to be an antidepressant as well: a doctor observed that patients with lung disease who were treated with this drug became noticeably cheerful. The antidepressant effects of imipramine were also discovered when it was used on patients.
- Hydergine is a drug in the ergot group that, due to the spectrum of its pharmacological effects, was first used therapeutically against hypertension and to stimulate peripheral circulation. In medical practice its alleviating effect on geriatric complaints was observed, and today it is used mostly in geriatric medicine.

- One further example: during the clinical tests of substituted arylsulfonyl-alkyl ureas in the tolbutamide class of drugs, which have blood-sugar-lowering activity, researchers noticed that they also have chemotherapeutic qualities.

It is this unpredictability—the impracticality of rationally designing chemical structures with specific pharmacological effects, let alone therapeutic effects; the impracticality of so-called drug design—that sets the boundaries for what sort of planning is possible in pharmaceutical research. Just as the possibilities for planning are limited by this unpredictability, however, the doors to chance are opened by it. When the accidental discoveries in pharmacological research are examined more carefully, though, it becomes apparent that they usually have nothing to do with pure chance but rather with what we call *serendipity.* The English author Horace Walpole added this word to our vocabulary in 1754 when he used it during a lecture he gave on the fairy tale "The Three Princes of Serendip." Serendip is the ancient name for Ceylon (Sri Lanka). The story is about three princes who discover—either by accident or sagacity—things other than those that they were seeking on their adventures. When imipramine was tested as a neuroleptic and Iproniazid as a tuberculosis drug and antidepressant effects were discovered as a result, this was not chance at work but *serendipity*—which requires attentive observation. To quote Louis Pasteur, "In the realm of scientific observation, luck is granted only to those who are prepared."

Now I have arrived at the second part of my essay, in which I will add my own personal experiences from my professional career.

EXAMPLES FROM SANDOZ RESEARCH

The operational plan laid out by Professor Arthur Stoll for his newly founded pharmaceutical department at Sandoz is an example of a research program based entirely on the first source of empirical knowledge mentioned earlier in this essay. The aim was to isolate the active principles of established medicinal plants in pure, intact form in order to provide the physician with a pure substance that could be measured and dosed. Stoll and his colleagues also developed new methods of isolation and purification for this purpose.

Two drugs that were utilized from the very start of the program were ergot (*Claviceps purpurea*) and squill (*Scilla maritima*). The ancient Egyptians used squill as a remedy for edema. There is evidence that midwives used ergot as a labor-inducing drug as early as the Middle Ages. At the time Stoll was doing his research contradictory assertions existed regarding the nature of the ingredients of these two drugs.

Ergotamine (Gynergen), a Natural Ergot Alkaloid

Stoll was soon able to isolate a crystallized alkaloid from the ergot; he named this ergotamine. Ergotamine was the first chemically pure alkaloid from this drug. An amorphous alkaloid preparation had already been isolated in 1906 by the Englishmen G. Barger and F. H. Carr; they called this ergotoxine, because its pharmacological attributes had a more toxic character. Ergotamine, on the other hand, demonstrated the active qualities of the entire drug: a contracting effect on the uterus and a sedative action on the sympathetic nervous system. Under the trade name Gynergen, ergotamine was introduced in gynecological medicine as a hemostatic for postpartum hemorrhaging and in internal medicine as a sympatholytic and sedative for the central nervous system. The ergot problem seemed to be solved. In 1936, however, the English gynecologist John Chassar Moir established that a water extract of ergot had a powerful uterotonic effect that could not be ascribed to the concentration of ergotamine, because this alkaloid is virtually water insoluble. Three years later this new, water-soluble ergot alkaloid that acted specifically on the uterus was isolated in four different laboratories. In England it was named "ergometrine." In Basel at the Sandoz laboratory, Stoll and Burckhardt isolated the compound that they named "ergobasine." Subsequently the International Pharmacopoeia Commission named it "ergonovine." At the Rockefeller Institute in New York, W. A. Jacobs and L. C. Craig discovered the structural core of the ergot alkaloids, which they named lysergic acid. They demonstrated that lysergic acid and the amino alcohol L-2-aminopropanol could be obtained through an alkaline hydrolysis of ergobasine. A complex structure was ascertained for ergotamine and ergotoxine, which showed how lysergic acid binds with a tripeptide-type residue.

At this time I had just finished my first substantial work at the Stoll laboratory: the structural elucidation of the *Scilla* glycoside. Because I was then available for a new assignment, Professor Stoll suggested I make a partial synthesis of ergobasine; in other words, a synthesis based on lysergic acid. This idea was not only of scientific interest; it had practical ramifications as well, because the medicinally valuable ergobasine was only present in very small amounts in the ergot, compared to the ergotamine and ergotoxine alkaloids. Professor Stoll warned me to expect difficulties due to the high volatility of the ergot alkaloids but was in agreement with the research project I had planned.

To obtain the lysergic acid necessary for the synthesis I could not use ergotamine but had to use the less expensive ergotoxine, which at that time was being extracted in larger quantities by Sandoz in their ergot-manufacturing laboratories.

During the purification of ergotoxine for the hydrolysis tests I made observations that led me to suspect that ergotoxine was not a pure alkaloid.

During tests aimed at the synthesis of ergobasine, I thus ran into a problem with ergotoxine. The solution to this problem later led to significant pharmaceutical preparations.

After initial difficulties that came about due to the great instability of lysergic acid, I was able to use the Curtius method as a procedure for synthesizing amides of lysergic acid by bonding with amines. The compound that resulted from the reaction of lysergic acid azide with L-2-aminopropanol was identical to ergobasine. We had thus achieved a partial synthesis of a natural ergot alkaloid for the first time.

A Half-Synthetic Ergot Derivative (Methergine)

With ergobasine as a prototype it was now possible to plan a research project based on this method of synthesis. The goal was the synthesis of oxytocic compounds—substances with a contracting effect on the uterus. In the framework of this project my colleague Dr. J. Peyer developed an economical process for synthesizing amino alcohol homologues, in other words, amino alcohols with different lengths of carbon chains that form amides with lysergic acid. In the pharmacology division at Sandoz, Professor Ernst Rothlin tested the chemical modifications of ergobasine obtained in this way using a special screening specifically targeted at testing the oxytocic effect. The next higher homologue of ergobasine—the modification with a side chain lengthened by one carbon atom, the d-Lysergic acid-L-butanolamide—showed optimal pharmacological qualities. This material, in maleate form, was sold for use in obstetric medicine under the brand name Methergine and is today one of the leading preparations for the treatment of postpartum hemorrhaging.

The development of Methergine from ergobasine is an example of making a derivative from a prototype with only minor alterations to the molecule. In a case of this sort it can be expected that the general pharmacological effects will remain and only quantitative differences will be measured. By contrast, if greater alterations in the structure of the prototype are made, then qualitative changes in the corresponding pharmacological effects can be expected. Nevertheless, even in this case chemists don't usually make modifications without a plan but rather base their work speculatively on structural precedents with known activity. When selecting which precedents to work from, the chemist's feelings or intuition can also come into play.

A Further Half-Synthetic Ergot Derivative (LSD)

Lysergic acid diethylamide, which I produced using the method of synthesis developed for ergobasine, was just such a speculative modification. Nicotinic acid amide was the model I worked from. It was sold under the brand name of Coramine and was a valued analeptic (circulatory and respiratory stimulant) in the pharmaceutical industry. The d-ring of lysergic acid is a modified nicotinic acid ring, and I had hoped—because of this structural similarity—to make a compound with a diethylamide that would also have an analeptic effect.

Professor Rothlin reported that during his research into the pharmacological effects of lysergic acid diethylamide, an oxytocic effect was recorded that was about 70 percent as strong as that of ergobasine. He noticed that the animals were restless during the narcotic slumber. The new compound with the laboratory name of LSD 25 (the 25th compound in the row of lysergic acid amides) met with no further interest by pharmacologists.

Intuitively at least, I had expected more from the pharmacological tests. Five years after the first synthesis I again made a small amount of LSD 25 in order to conduct further tests on the compound. As with the first occasion, the amount manufactured was only a few hundredths of a gram. The synthesized product had to be separated from the isolysergic acid isomers on the Alox (aluminum oxide) column; for this we used dichloroethylene as a solvent. Then the lysergic acid diethylamide was crystallized out of methanol as a tartaric acid salt.

On that afternoon while I was engaged with

this work I fell into a strange, dreamlike state. I went home early and had the urge to lie down when I got there. I did so, and when I shut my eyes I experienced a fantastic play of color and forms. Whatever I thought to myself appeared in vivid imagery before my inner eye. This unusual state of consciousness, which was not unpleasant, vanished after a few hours.

I suspected that I had been intoxicated by something in my laboratory. At first I thought the cause might be the dichloroethylene that I had used in the chromatography. The next day at work I cautiously inhaled some vapor from this solvent. But nothing happened. Then I tested the lysergic acid diethylamide that I had also worked with on the day of the incident. But because I had been habitually working in a clean environment with the ergot alkaloids I could not actually imagine how I could have possibly taken enough of this substance for it to have any effect. If LSD 25 was indeed the cause of this disturbance, then we must be dealing with an extraordinarily active substance. Being a cautious man, I began my self-experimentation with the suspect substance using the smallest amount from which an effect might be expected, namely 0.25 milligram of lysergic acid diethylamide tartrate. I figured I would cautiously increase the dosage from there. But I never got that far. The first dose of a quarter milligram caused a state of intoxication that lasted around twelve hours. The profound, radically shifting transformations that I experienced of both the external world and my own consciousness have been described often enough and need not be repeated here. This first planned LSD experiment was an unusually frightening experience, because I didn't know if I would ever return to everyday reality and to my normal state of consciousness. Only as I began to sense the slow return of a familiar reality was I able to enjoy the exquisitely stimulated visionary experience.

With the discovery of LSD a psychoactive substance of as-yet-unknown potency had been found. I had been looking for a circulatory stimulant and discovered a psychological stimulant: *serendipity.* Despite my caution, in this first planned LSD experiment I had selected an amount that was five times larger than the average effective dose of about 0.05 mg. In order to get a sense of the potency of LSD, just imagine that a gram of it is enough to put about twenty thousand people into a state of hallucinogenic intoxication.

The highly specific effect of LSD made this substance into a valuable tool in psychiatric and neurophysiological research. LSD also found use as medicinal aid in psychoanalysis and psychotherapy.

But LSD research experienced a serious setback, and one from which it has yet to recover, when the substance became swept up amid a general wave of intoxicants in the USA. For a period of time in the mid-1960s it made headlines as the "No. 1 drug." Its misuse—in other words, the way it was used by the drug scene—bore no relationship to the actual effect and character of LSD. In many instances this led to serious accidents, and gravely damaged further LSD research.

In my opinion, however, the last word has not yet been spoken about the possibility of medical applications for this effective substance.

I will now discuss an aspect of pharmaceutical research that is rarely talked about in scientific publications. This is the fact that a researcher's secondary paths of investigation often lead to more significant results than the primary paths that were established on the basis of planning. Such a secondary path of investigation might be taken when a chemist follows up on so-called chance observations, or is able to follow up on them—something that is not usually allowed in strictly targeted research projects.

I hit upon such a fertile secondary path myself when I pursued observations I made during the purification of the initial materials for the synthesis of ergobasine, specifically during the purification of ergotoxine. As previously mentioned, I had the impression at the time that ergotoxine was not a single alkaloid. In fact, ergotoxine can be separated into three different alkaloids by using an acid specifically made for salt formation with lysergic acid derivatives. One was identical to ergocristine, the alkaloid that had been isolated a short time before by Stoll and Burckhardt in the Sandoz ergot-manufacturing department. The other two alkaloids were new. I named one ergocornine and the other ergocryptine (because it had remained hidden longer in the mother liquor). Many years later it was shown that there are two isomers of ergocryptine; we call these alpha-ergocryptine and beta-ergocryptine.

Hydrogenated Ergot Alkaloid (Hydergine and Dihydergot)

At this point ergotoxine, which had been considered a single alkaloid for thirty years, had been divided into three—or rather, four—further discrete alkaloids. Now pharmacology tests could proceed on a firm basis. It was apparent that the contradictory reports found in the literature about the effects of ergotoxine could be traced back to the fact that the different researchers were working with different ergotoxine preparations, and it was discovered that the ratios of the active ingredients in ergotoxine preparations varied widely between different sources. I provided Professor Rothlin with not only specific ergotoxine components for pharmacological testing but also their dihydro-derivatives. Jacobs and Craig of the Rockefeller Institute had shown that the exocyclic double bond of lysergic acid can be selectively hydrogenated and that dihydro-lysergic acid is stable, while lysergic acid and its derivatives form an equilibrium between lysergic acid and isolysergic acid when put in a solution. The isolysergic acid derivative proved to be much less pharmacologically active than the corresponding lysergic acid derivative. Professor Rothlin and his colleagues in the pharmacology department found that the ergotoxine alkaloids stabilized by hydrogenation—dihydroergocristine and dihydroergokryptine—presented an interesting method of action that was markedly different from that of the natural, nonhydrogenated alkaloids. Instead of vasoconstriction, vasodilation, and blood-pressure-lowering effects, they demonstrated increased sympatholytic characteristics and a considerably reduced toxicity. Because of this profile of pharmacological effects, the dihydro-derivatives of the three ergotoxine alkaloids were combined in a 1:1:1 ratio in the form of their water-soluble methanesulfonates and introduced as a therapeutic drug under the brand name Hydergine. As previously mentioned, Hydergine was not retained as an antihypertensive but proved in practice to be an effective geriatric medicine and remains one of the most important of the Sandoz pharmaceutical products.

At that time I had also hydrogenated ergotamine and introduced dihydroergotamine into the pharmacological tests. The intention was only to obtain a stable ergotamine, not a pharmacologically different derivative of ergotamine. Ergotamine preparations, such as were marketed in the form of ergotamine tartrate under the name Gynergen, had the disadvantage that a higher percentage of the ergotamine they contain changes into the pharmacologically inactive isolysergic acid alkaloid ergotarminine. We hoped to remedy this by using hydrogenation, which has a stabilizing effect. The hydrogenation not only had a stabilizing effect but also caused a profound change to the profile of pharmacological effects, just as it had with the alkaloids of the ergotoxine group. Compared with ergotamine and its vasoconstrictive effect, dihydroergotamine instead demonstrated a stabilizing effect on the blood pressure and the vascular system, a further strengthened sympatholysis, and less toxicity. This profile of pharmacological effects proved to be therapeutically valuable. Dihydroergotamine is now used in therapy under the brand name Dihydergot for the treatment of orthostatic hypotony and vascular migraines. We had planned the stabilization of ergotamine—we found a new medicine.

This brings me again back to LSD. Even if LSD itself had not become a therapeutically useful pharmaceutical product, it was the impetus for the development of a new medicine and indirectly led to the development of a further pharmaceutical preparation.

THE SERETONIN ANTAGONIST LSD LEADS TO DESERIL

The English physiologist Sir John Gaddum (1900–1965) established that LSD is a highly active serotonin antagonist. Serotonin is a broadly distributed endogenous chemical found in mammals that also plays a role in the psychological functions of biochemistry. Because serotonin is involved in the inflammation process as well as with certain forms of migraines, a serotonin inhibitor might possibly exhibit therapeutic characteristics. But LSD as such could not be used therapeutically as a serotonin inhibitor because of its psychological and hallucinogenic characteristics. Our pharmacologists therefore suggested that we search for chemical modifications of LSD that might still have the effect of being a serotonin antagonist but no longer have the hallucinogenic characteristics. Of the many LSD derivatives formulated by my colleague Dr. Franz Troxler, Brom-LSD proved to be the first serotonin antagonist without a hallucinogenic effect. With the expanded testing of other lysergic acid derivatives, an optimal serotonin antagonist was finally found in 1-methyl-d-lysergic acid-(1-hydroxybut-2-yl)amide. The preparation was introduced into therapeutic use for the interval treatment of migraines under the brand name of Deseril (Sansert).

As a conclusion I want to talk briefly about further developments at Sandoz that came about in the wake of the LSD research.

FROM LSD TO THE MEXICAN MAGIC MUSHROOMS

In the spring of 1957 Dr. Yves Dunant, at the time the director of the Sandoz branch in Paris, relayed an inquiry from Professor Roger Heim, director of the Laboratoire de Cryptogamie in Paris, asking whether we would like to participate in the chemical investigation of Mexican magic mushrooms (*Psilocybe mexicana*). I happily agreed. These mushrooms were used by the native people of Mexico in religious and ritual contexts as well as for medical practices with a magical orientation. Professor Heim had already classified them botanically. He was also able to produce a number of new mushroom species in the laboratory, most of them in the *Psilocybe* genus. The ancient secret mushroom cult in Mexico had been rediscovered in 1954 and 1955 by the American research team of R. Gordon Wasson and his wife, Dr. Valentina Pavlovna Wasson. Previous attempts in Paris and in two other laboratories in the USA to isolate the active ingredients in the mushrooms had proved unsuccessful, so Professor Heim turned to us. He believed we might be in a better position to solve the problem due to our experiences working with LSD, which qualitatively displayed the same effects as the magic mushrooms.

None of my colleagues at the time were willing to take on the investigation of the mushrooms, because in those days anything that was connected at all with LSD was viewed unfavorably by the top management. Therefore I took on the chemical-isolation investigations myself, together with my efficient laboratory assistant of many years, Hans Tscherter.

My colleagues in microbiology at the time, Drs. Artur Brack and Hans Kobel, were able to significantly improve the laboratory culture of *Psilocybe mexicana.* With this mushroom material and the testing of the extracts through self-experimentation, in which many of my colleagues and peers took part as guinea pigs—for the animal tests had provided no clear results—we were able to isolate the active principles and crystallize them in pure form. We named them psilocybin and psilocin.

With the purified active ingredients at hand, my colleagues Drs. A. J. Frey, H. Ott, T. Petrzilka, and F. Troxler joined forces, and we were able to clarify the structure and achieve the synthesis.

The structure of the mushroom's active ingredient is noteworthy among other things, because it is closely related to serotonin as well as because, like lysergic acid and LSD, it incorporates an indole derivative that is substituted at position 4. In my opinion psilocybin and psilocin deserve further attention in experimental psychiatry and as prototypes for chemical modification. But psilocybin research has already produced useful information and led indirectly, as LSD did, to important new medicines.

From Mexican Magic Mushrooms to Visken

Dr. Troxler, following his development of an efficient synthesis of the basic material for psilocybin, 4-hydroxyindole, was engaged in a project aimed at the development of a new substance with an inhibiting effect on the adrenergic beta-receptors. Beta-receptor inhibitors (beta blockers) are used therapeutically for the regulation of heart function. A well-known prototype with this effect was propranolol (Inderal). It had been established that substances with this type of effect are created when the isopropylamino-2-hydroxypropyl side chain characteristic of propranolol forms an etherlike bond with an aromatic system of phenolic hydroxyl groups. Among the many phenols Dr. Troxler used for synthesizing the substances he would test as beta-receptor inhibitors, one was the rare 4-hydroxyindole, which he had available to him although it could scarcely be found in any other laboratory in the world. And it was exactly this combination that proved to be a winner. Under the brand name Visken, the active ingredient gained a leading position among the beta blockers, especially for the treatment of hypertonia.

Without LSD, the magic mushrooms would never have ended up in our laboratory; without the work on the magic mushrooms, 4-hydroxyindole would not have become available, and Visiken would not exist.

With this I come to the end of my discussion. I hope to have shown how pharmaceutical and chemical research does not and cannot always proceed on quite as straight a path as the highly stylized articles found in scientific journals might like to make it seem. Rather it is chance—or better, what Walpole called *serendipity*—that often contributes a great deal to success and which will certainly continue to contribute to accomplishments in the future.

REFERENCES

More information from the original publications about the ergot alkaloids and their derivatives can be found in the following monographs.

Berde, Botond, and Heinz Otto Schild. *Ergot Alkaloids and Related Compounds.* Berlin, Heidelberg, and New York: Springer-Verlag, 1978.

Hofmann, Albert. *Die Mutterkornalkaloide.* Stuttgart: Ferdinand Enke Verlag, 1964. Reprinted Solothurn, Switzerland: Nachtschatten Verlag, 2000.

Hofmann, Albert, Roger Heim, Arthur Brack, Hans Kobel, Albert Frey, Hans Ott, Theodor Petrzilka, and

Franz Troxler. "Psilocybin und Psilocin, zwei psychotrope Wirkstoffe aus mexikanischen Zauberpilzen." *Helvetica Chimica Acta* 42 (1959): 1557–72. Hofmann's article about the Mexican magic mushroom (*Psilocybe mexicana*) research.

If a chemist does not become a mystic,
then he is not a chemist either.

Can Insight into the Truths of Natural Science Be Therapeutically Effective in Psychology?

This lecture was presented at the Fourth Symposium for Consciousness Studies, sponsored by the European College for Consciousness Studies (Europäisches Collegium für Bewusstseinsstudien; ECBS), December 8–10, 1989, in Freiburg im Breisgau, Germany.

If my answer to this question is yes, then it is based on personal experience. Because I want to talk about my personal experience I will dare to express my own thoughts about a topic that has its proper place in psychotherapy. Because I am a chemist this means I am stepping outside the boundaries of my discipline.

The following statements, which describe how scientific insights can have psychotherapeutic effects, can be divided into two parts.

In the first part I will describe how meditating on scientific knowledge helped to alleviate a phase of severe depression in my life.

In the second, more general part, I will attempt to present my view of how an increased awareness of the truths of the natural sciences could contribute a great deal toward helping people to overcome their spiritual crises and to reduce their fear, depression, and loneliness.

Now for my own personal experience. In my twenty-ninth year I fell into a severe existential and spiritual crisis. The external world became unreal and increasingly lost all meaning. Everything going on seemed pointless to me. My fellow inhabitants moved around like wooden puppets. I was completely overcome with a fear of dying. Often these symptoms were so intense that I collapsed. At night I was afraid of falling asleep, because I feared I would never wake up again. I felt guilty about the state I was in, but I could not determine what palpable sin I was burdened with. My soul was in a terrible state. Because I believed my suffering was of a purely mental sort, I tried to get out of this dire situation by applying my reason, through self-analysis, and with the full force of my will. But all attempts were futile, and my fear only increased.

As I was brooding one day in my room, looking out the open window, my gaze fell on a green tree in the garden. A curious relationship effortlessly arose with this tree, which penetrated the vicious circle of muddled thoughts in my head and the path to healing was cleared. The following meditation went through my head: This tree is built in the same biochemical manner as you are; it is made up of cells with a nucleus that contains the hereditary factors and which is surrounded by a protective plasma hull, just like the cells of your body. It has come into being through the union of a female and a male cell, just like you. It develops and grows by breathing the same air as you; it is fashioned and kept alive by the same creative force as you are.

The understanding that flooded my consciousness about my shared creation with this tree, which apparently lived its life unconstrained by thoughts, suddenly filled me with serenity and trust. The mental commotion and anxiety disappeared.

When anxieties later began to resurface I only had to think of brother tree in his serenity in order to feel protected again and supported by a shared generative spirit. It is conceivable that even with no knowledge of the natural sciences we could experience our shared creation with this tree, and consequently with the entire plant world, as an element of healing, as a spontaneous vision, such as St. Francis of Assisi experienced when he talked with the trees and animals. In my own childhood I often had such deeply uplifting experiences during moments when nature, the forest, or a meadow of flowers suddenly appeared to me in a clear light, in meaningful beauty, and gave me a feeling of blessed security.

Such a visionary experience of a more profound and uplifting reality is in no way rare for children—this is what is meant by the "paradise of childhood."

But with adults such spontaneous mystical visions are rare—which is why in my case I believe it was the knowledge and contemplation of the scientific truth concerning man's shared creation with the tree that triggered the healing experience.

I now come to the second and more general part of my observations. At this point I want to include insights from the natural sciences—insights that I believe deserve to be reawakened in the consciousness of a greater number of people and to be broadly disseminated in the public sphere. These insights have nothing to do with new research. They are scientific facts that can be read about in schoolbooks, although they are not given further consideration because they have no direct relationship to everyday life and are of no immediate use.

At this point I would like to dispense with any fundamental ideology regarding the position and significance of research in the natural sciences and about the truths arrived at through this research. I do not believe that the significance of the natural sciences in the evolution of human society rests solely on the fact that they deliver technological developments in modern industry, which have fundamentally changed our life and our planet. The significance of the natural sciences lies in their ability to open people's eyes to the wonder of Creation and to the unity of all life on this Earth, humanity included. Gaining such a broad and holistic consciousness could become the foundation for a new spirituality. It could contribute solutions to current spiritual, social, and ecological problems.

The Indian philosopher Rabindranath Tagore expressed this belief with poetic language.

> Through our progress in science the wholeness of the world and our oneness with it is becoming clearer to our mind. When this perception of the perfection of unity is not merely intellectual, when it opens out our whole being into a luminous consciousness of the all, then it becomes a radiant joy, an overspreading love.*

*Rabindranath Tagore, *Sādhanā: The Realization of Life* (New York: Macmillan, 1913), 113.

Before I come to the examples of which truths in particular from the natural sciences should enter into general consciousness, I would like to define what is meant by *truths of the natural sciences.*

What is truth? Pontius Pilate once asked this question, and he washed his hands of guilt. But today we know a little more. Today this question is often asked by philosophers, and especially by psychologists, and it is left open in the same sense as it was with Pilate, namely: we do not know what truth is. To put it another way, there is no truth, everything is relative, it is all just illusory. This fatalist position, which has resulted in a perceived relativization of everything and everybody, is one of the negative hallmarks of our times. It is a source of psychological instability, purposelessness, fear, and other mental afflictions.

The assertion that there is no truth, that everything is relative, that it is all just illusory, is a demoralizing worldview that has had devastating consequences. These can hardly be overstated. It originates in a type of thinking that is incapable of differentiating between the subjective and the objective portions of the all-encompassing, true reality.

In the following remarks I would like to try to clarify the fundamentally important difference between scientific, objective truth and truth that cannot be objectified—subjective truth.

This relates to the question of what happens in our experience of life in the external world and what happens in our internal world. What is really outside and what is inside? It is obvious that for human experience both are important—an external material world and an experiential human individual world.

In order to demonstrate the necessary reciprocal relationship between the external material world and the internal mental world of human beings, I will use the example of how television signals of image and sound are transmitted through the airwaves to a television set.

The material world in the external realm acts as the transmitter, sending optical and acoustic waves, delivering sensory, gustatory, and olfactory signals to the receiver. The receiver is in the mind of the individual person, where the sensory organs—the antennae—transform the stimuli into a sensual and mental picture of the outside world.

If one or the other were missing—the sender or the receiver—then there would be no human reality, just as the television screen would remain empty and silent.

The following will show what we know about the human experience of reality based on insights made by the natural sciences into human physiology with regards to our function as receivers. We will also look at the mechanisms of reception and experience.

The antennae of the human receiver are our five sensory organs. The antenna for optical images from the external world, the eye, is able to receive electromagnetic waves and produce images out of them on the retina that correspond to the object from which these waves emanate. From there the nerve impulses corresponding to the image travel via the optic nerve to the optic center of the brain, where these impulses (which up to that point are objective electrophysiological, energetic events) become the subjective mental phenomenon of seeing.

It is important to keep in mind that our eye and our inner mental "picture screen" use only a very small section of the enormous spectrum of possible electromagnetic waves in order to make the outer world visible. While the electromagnetic wave spectrum encompasses wavelengths that range from a billionth of a millimeter (as in x-rays), to radio waves of many meters long, our ocular apparatus recognizes only a very narrow section of wavelengths—between 0.4 and 0.7 thousandths of a millimeter. Only this very limited section of the entire spectrum can be received by our eyes and understood as light. Within this narrow section of visible waves humans are capable of perceiving and distinguishing the various wavelengths between 0.4 and 0.7 as different colors.

It is important to keep in mind that there are no colors in the external world. People are not generally conscious of this fundamental fact. What we see as color in an object that exists in the outer world is exclusively material that transmits electromagnetic waves of certain lengths. When an object that is struck by light reflects back waves of 0.4 thousandth of a millimeter, then we say it is blue. But if the object sends waves of 0.7, then we describe this optical experience as red. It is impossible to determine if all people have the same experience of color when they perceive the same electromagnetic wavelength.

The perception of color is a purely mental and subjective event that takes place in the inner realm of the individual. The world of color, as we have seen, does not exist objectively in the outer world but is created on the mental picture screen in the interior of the individual human being.

In the auditory realm corresponding relationships exist between the senders in the outer world and the receivers in the internal one. The antenna for acoustic signals, the ear, in its function as a part of the human receiver, only has a limited range of reception. As with colors, sound does not exist objectively. What is objective in the process of hearing is, again, waves: wavelike compressions and rarefaction of the air are registered by the eardrum in the ear and transformed into the mental experience of sounds in the primary auditory cortex of the brain. Our personal receiver for acoustic waves responds to those in the range between 20 oscillations per second and 20,000 oscillations per second (which correspond, respectively, to the lowest and the highest sounds perceptible by human beings).

Other aspects of reality that are perceived by the remaining three senses of taste, smell, and touch are also the results of a reciprocal interaction between material and energetic senders in the outer world and the mental receivers in the inner realm of the individual human being. I need not describe them here in detail. This information can be found in any physiology textbook. It should only be noted that the senses of taste, smell, and touch—just like color and sound—cannot be determined objectively. They exist only on the mental picture screen in the internal realm of the individual human being.

From these considerations it follows that the world as we perceive it with our eyes and other sensory organs is a reality uniquely tailored for humans. It is determined by the capabilities and limitations of the human sensory perception. Animals with different sensory organs, with antennae that respond to different types of waves and wavelengths than humans, see and experience the external world in a completely different way. They live in a different reality.

The metaphor of reality as the product of a sender and receiver demonstrates that the seemingly *objective* picture of the external world—what we call "reality"—is actually a subjective image. This basic fact shows that the picture screen does not exist in the outer world but only in the inner realm of every human. All individual human beings carry their own, personal picture of reality—a picture created by their own private receptors—*inside* themselves.

Every human being has their *own* true worldview that can be perceived with their own eyes and other sensory organs.

But is there a part of these individual worldviews that might be true in a sense that is objectively valid for all human beings? It is the material that can be objectively confirmed and measured by means of the scientific method. It is what we have referred to as the "transmitter or sender."

There is only *one* sender. There is only *one*

scientific reality, but there are as many subjective worldviews as there are receivers, in other words, individual human beings.

Now I would like to talk about the great transmitter that sends such objective data that is true for all human beings. What is it made of? It consists of matter and energy and nothing more: matter, which is characterized by its chemical and physical qualities, and energy, which comes in the forms of radiation and thermal, electrical, and mechanical energy.

Everything emanating from the sender that can be objectified and measured in the external world is always related to matter and energy. This is the area of research in the natural sciences. But what the sender is able to reveal to the receiver solely by means of its material and energetic signals is phenomenally endless, and it is wonderfully inexplicable how our receiver is able to transform these signals into a lively and colorful world.

Let's begin with the chemical makeup of the sender. The progress of the natural sciences from antiquity until modern times has provided us with surprising insights. In place of the four elements of earth, water, fire, and air, ninety-two more have been discovered and arranged in a numeric system: the periodic table of elements. A number of unstable radioactive elements have also been added. Beyond this we have gained insight into the inner construction of atoms, where a microcosm was discovered—a nucleus encircled by electrons like the planets around the sun. But research did not end with the atomic nucleus. This itself is made up of protons and neutrons. These too can be divided into a multitude of elementary particles with extremely short life spans, which finally dissolve into waves and transform into energy.

Chemical and physical research has further established that the senders—the universe, the sun and the planets, the creations of this Earth, as well as our bodies—are all built from the selfsame primordial material. Mystics also experience this scientific truth when they feel themselves to be physically unified with the universe.

The natural sciences have also provided us with profound insights into the macrocosm. To a precise degree of accuracy we know the movement of the planets that have orbited the sun from time immemorial. Stars, the siblings of our sun, have been studied, and our solar system has been measured in light-years. Other galaxies have been discovered millions of light-years away. The "eyes" of the radio telescope continually probe deeper into space, discovering new galaxies and revealing the infinity of the universe.

Astronomical research has provided us with photographic images of our Earth, a blue ball floating in space. Research in physics, chemistry, and biology has brought the uniqueness of our spaceship Earth into focus. The relationship of Earth to the sun has been clarified, with the sun as a giant atomic reactor, delivering inexhaustible energy for all the processes of life on Earth. With sunlight as the source of energy, the green plants that cover the Earth are able to make an organic substance, our food, from the inorganic matter of carbon dioxide and water. This food contains stored energy from the sun. Through digestion, the process is reversed in our bodies, again generating carbon dioxide and water while the stored energy is released. Energy from the sun builds up our bodies, and energy from the sun is what keeps the process of life going. Even the brain's functions are nourished by the sun's energy, and thus the human mind represents the most sublime energetic transformation of sunlight. In other words, research in the natural sciences has revealed that we human beings are "sun beings"—a truth represented in many myths.

These few examples must suffice in order to give us an awareness of the inexhaustible scientific aspects of the great sender. It is a tiny section of the objective reality of the world we live in. Such sections are passages in what Paracelsus called the "book of nature," which was written by the finger of God, and in which we should learn to read. Indeed, it contains a first-hand revelation of the truths of natural law. The infinity of the starry sky and the beauty and fertility of the Earth, with all of its wonderful creatures of the animal and plant kingdoms, are part of this revelation. Those who live in harmony with Nature and her rules will remain physically and spiritually healthy. This is true for individuals and for all of humanity.

The looming environmental catastrophe clearly demonstrates the consequences of acting against the laws of nature.

That is enough about the outer world for the moment. This sender makes up half of human reality. Now I want to look more closely at the other half, the receiver. It has already been noted that there is only one sender, but there are as many receivers as there are individuals. There is no collective receiver. Only the individual can see, hear, feel, and perceive. The individual is the receiver, or rather, it is the consciousness of the individual that is the receiver, for the body is a material object in the external world and therefore is part of the sender. I can see my body, and I can also experience it with my other senses. My senses, the antennae of the "receiver-self," are likewise part of the external world. This is not only true with regard to the eyes and ears but also applies to the neural pathways that lead to the brain: they are material, as is the brain itself. The electrical currents and impulses that travel as signals from the external world into the brain, and carry on their activity within the brain, are to be classified as objective, energetic phenomena. But at this point there comes a great gap in human knowledge: the transition of material-energetic occurrences into an immaterial, mental, and spiritual picture that is no longer objective, in other words, the transition into subjective perception and personal experience. At the same time this gap in knowledge is the interface between sender and receiver: the place where they merge into one another and combine to complete a living being.

We now return to the question of the receiver, and to the question of consciousness. What is consciousness? Consciousness eludes any scientific explanation because it would, of course, be my consciousness explaining what consciousness is. Thus, all attempts at clarification remain tautologies. Consciousness can only be paraphrased as the receptive and creative spiritual center of the individual, and it can only be described by its contents and its abilities.

In regard to this central question I would like to quote a definition of consciousness from Jean-Paul

Sartre: "Consciousness is the absolute, the I; it is an individual's transpersonal dimension of being." This seems to me to be yet another tautology.

Personal consciousness could be understood as an individual construction of the greater mental-spiritual universe, just as our bodies are scientifically verifiable constructions of the material universe.

Both the material universe and the mental world are infinite. Incorporated in both, the human being takes part in their infinitude.

The metaphor of sender and receiver appears to correspond to a dualistic concept of the world consisting of a material outer world and mental inner world. But just as the image and sound of television only exist when there is a transmitter as well as a receiver, so does human reality only exist as a unified whole of mind and matter. The sender and receiver are merely a construct of our intellect, necessary for a rational description of the reciprocal relationship between the inner and outer world that make up and effect reality.

Reality is a result of this relationship, and the

sender-receiver concept points to the basic fact that reality is not a firmly defined state. It is not static but is the result of an ongoing process: a continuous input of material and energetic signals from the external world and its continuous decipherment—in other words its transformation into mental experience—in the internal realm. Reality is a dynamic process; it is new in every moment.

Thus, actual reality only exists in the here-and-now, in the present moment. This is why a child, who lives more fully "in the moment" than an adult, is able to experience a more genuine picture of the world. A child lives in a world that is open to more reality, more truth.

The momentary experience of true reality is a primary concern of the mystic. This is the point where childhood experience and mystical experience meet.

If this understanding of the dynamic nature of true reality is meditated upon, it might help to provoke a spontaneous mystical experience. Mystical experiences are recognized as a healing factor in the case of many mental ailments. Thus, insight into scientific truths concerning the dynamic character of reality may also have a psychotherapeutic effect.

The concept of reality as the product of a transmitter and receiver is particularly meaningful with regard to the role of the receiver, the individual, in the construction of reality. This concept reveals the creative potential bestowed upon each individual. It makes us conscious of the fact that every human being is the creator of his or her own world. For it is only inside of ourselves that the Earth and the colorful life upon it come alive, that the heavens and the stars become a reality.

It is in this truly cosmogonic skill, this ability to create our own worlds, where the actual freedom and responsibility of every individual is to be found.

When I am able to recognize the aspect of reality that is objectively external and the aspect that is happening subjectively inside me, then on the one hand I know better what I am able to change in my own life, where I have a choice, and what I am responsible for. On the other hand, I know what lies outside of my personal will and what must be accepted as an unchangeable fact.

This clarification of my responsibility in the creation of reality is an invaluable aid in my life. It offers an insight into conditions of our human existence that are verifiable by the natural sciences—and knowledge of these facts can have a psychotherapeutic effect.

LIGHT—LOVE—LIFE

Evolution in three words:
Light of the sun is primordial energy,
it wakes and maintains
Life on earth,
the highest development of which
is Love.

MEDITATION AND SENSORY PERCEPTION

The Search for Happiness and Meaning

This lecture was presented at the annual meeting of the European College for Consciousness Studies on May 30, 1997, in Leipzig, Germany.

Meditation is defined in the Kröner *Philosophisches Wörterbuch* (Philosophical Dictionary) as "reflection, contemplation, and observation in the philosophical and metaphysical sense. In the religio-mystical sense, meditation is experienced as an immersion, as the means to the most profound understanding."

What is it that we reflect upon, contemplate, and observe? What is it that we wish to understand in the most profound way? It is the content of our own consciousness. This content was brought in by means of sensory perception. Sensory perception precedes meditation.

Why do we meditate? What purpose does it serve? There must be a purpose and it must have a meaning, otherwise no one would do it. It could be said that by meditating one seeks new aspects or new depths of reality, or that one strives to know oneself better, or tries to comprehend a particular experience. There are infinite phenomena, concrete and abstract, which could be the object of meditation. But is there also a common denominator that includes all of the different types and goals of meditation? What lies at the root of all of this searching through meditation?

While I was pondering this question in preparation for today's lecture I received an inquiry from the directors of this year's Basel Psychotherapy Symposium asking if I was willing to present the opening speech at the conference. I agreed, because the congress was devoted to the theme of "The Search for Happiness and Meaning." For ultimately, the search for happiness and meaning is not just the concern of psychotherapists, but also of chemists. It is, perhaps, the primary concern of all human beings.

With this, I had the answer to my question, "What lies at the core of the searching through meditation?" It is the search for happiness and meaning.

What just occurred was a typical case of synchronicity: two unconnected causes led to a complete and meaningful event.

"Meditation and sensory perception" and "the search for happiness and meaning" contain the same basic themes.

If we mean the "search for meaning" in the broadest sense, in other words, the meaning of human existence, then we can now clarify this meaning.

Basically all great religions and philosophies arise from the search for the meaning of creation and of human existence, and they also offer an answer to this all-encompassing question.

Although the answers that are given vary greatly from one another, each contains a promise of happiness: happiness of the eternal soul in the Christian heaven; the happiness of sensual pleasures in the Paradise of Islam; or the earthly happiness of the Epicureans.

More than two thousand years ago, at the beginning of his *Nicomachean Ethics,* Aristotle asked the question, "What do humans seek?" And he discovered that their loftiest ambition and their greatest possession is happiness.

Thomas Aquinas arrived at the same answer when considering the question of the purpose of human existence. He formulated this in the famous sentence, "*Ultima ratio vitae humanae beatitude est*": the ultimate meaning of human life is happiness.

When we come to the philosophers of the modern era, their work is also ultimately about the search for happiness and meaning. I shall quote only one modern philosopher, Ludwig Marcuse, who reaches the following conclusion in the introduction to his book, *Philosophie des Glücks* (Philosophy of Happiness): "He who denies himself happiness does not fulfill his reason for existence."

What the founders of religions and philosophers say about the ultimate meaning of our existence—that happiness is the meaning and the final goal in life—must be true, because no one could agree with the opposite, that the purpose of our life is to be unhappy.

What is happiness? Philosophers have debated this since antiquity, and the discussion continues today. Books are written and symposiums are held which address the question "What is happiness?" The search for happiness endures.

There are two places where humans seek happiness: in the realm of being or in that of having, which is to say, on the spiritual plane or the material one.

Nowadays, and especially in the Western world, happiness is hectically sought in the material realm—with variable success. Today there are more and more people who are rich and super-rich, but scarcely more happy people. On the other side, there are more and more people who are poor, some living in misery, who are extremely unhappy.

In Aldous Huxley's final book, *Island,* a wise regent on this happy utopian island made a rule in order to prevent such an accumulation of money and wealth by forbidding that any inhabitant earn more than three times as much as the average. Huxley had foreseen the catastrophic development of the modern world.

What is the purpose of the accumulation of money and power by individuals or corporations that have no responsibility to the public good? The responsibility for public welfare lies with the state, for example in the care of the unemployed. But the state has no power in the economy, upon which the welfare of its citizens is critically dependent. Responsibility and power have drifted apart from one another with catastrophic consequences.

In the context of the search for meaning and purpose it is worth considering the difference between *possessions,* in the original sense of the word, and *property.* This might help to bring the tragic absurdity of contemporary developments into clearer focus.

Words are created from an immediate experience of reality and are related to the basic facts and activities of our existence. When the word *possession* [*Besitz*] came into use by early human communities, it meant merely a thing of personal use, for example a horse that one sat on, or the chair upon which one sits.*

Since then the word *possession* and the verb *to possess* [*besitzen*] have taken on much broader and more symbolic meanings.

In a later period the legal term *property* [*Eigentum*] came into use.† Property signifies the legal recognition and protection of possessions. Today the two words are considered synonyms and are used interchangeably. But the fact that they originally meant something considerably different is evident in the fact that "possession" has a related transitive verb, "to possess," but with the word "property" no corresponding verb exists.

---

*[In the German, these examples are revealing in a literal way, because the noun *Besitz* is related to the verb *sitzen,* "to sit." —*Trans.*]

†[In German, the word *Eigentum* is related to the verb *eigen,* "to own," and has a connotation of legal ownership. —*Trans.*]

Since the introduction of the legal term *property,* it has become possible to have more "possessions" than one can actually "own" in the original sense of the word—in other words more than one can personally use. The advent of this possibility planted the seed for a significant part of human tragedy that has been created by the accumulation of money and power.

If this differentiation could be made conscious, if more people would strive toward true ownership rather than property, then a lot of fruitless endeavor, conflict, and unhappiness would disappear, and a corresponding equanimity, joy, and happiness would replace them.

What I am referring to here finds its pithiest expression in a Chinese aphorism: "The landlord says, 'my garden . . . ,' and his gardener laughs."

The landlord has the right to refer to it as his garden, because it is his property. But perhaps he is rarely seen there. Or maybe he takes a stroll through it once in a while and shows his visitors this or that particularly beautiful plant and the newly built pavilion, but he does not have any deeper emotional connection to his garden. For his gardener, on the other hand, this garden is a living thing. He lives with it, and it lives with him. He planted the trees, he prepared the vegetable beds—he knows every single flower, every branch. He cares for them with love; he observes their growth, their blossoming, and their decay. He knows the garden in the morning dew; he goes through the beds again in the evenings when some flowers more powerfully exude their scent; and in the afternoon heat he likes to nap in the pavilion during his break. He loves this garden with his whole heart. It is he who possesses the garden from early in the morning until late in the evening. He is its true owner, and so he laughs when the landlord says, "My garden . . ."

As this example of landlord and gardener, of owner and possessor, demonstrates, one needn't be the owner of the meadows, fields, and forests through which one wanders in order to enjoy the flowers that grow at the edge of the path or to appreciate the intoxication of the trees and of everything else that otherwise offers such an escape for the eyes and ears.

Let us return to our theme of "the search for happiness and meaning."

When one seeks something, one should actually know what one is searching for. But happiness cannot be scientifically defined. It is a conclusion, so there is no further explanation. Happiness can only be paraphrased as a particular state of human consciousness. Happiness belongs in the category of being. It is not, therefore, something one can have. What is sought in the search for happiness is, in reality, not the happiness itself but rather what we believe or hope will make us happy. In truth and in deed, the search for happiness is a search for the source of happiness.

What the source of happiness might be or should be has been debated since antiquity. St. Augustine mentions that 288 different doctrines concerning the subject of happiness were recorded in a Roman encyclopedia.

Regarding the views on what happiness is, or better, what makes a person happy, I only wish to mention one philosopher of our age, Friedrich Nietzsche, who was not happy himself but who thought profoundly about the nature of human existence and about happiness. He wrote, "The happiness of human beings rests upon the idea that there is an indisputable truth."

Earlier, in perhaps happier times, the dogmas of the Church were considered indisputable truths. Today it is the findings of the natural sciences that are considered indisputable truths and that have made the ancient religious worldview implausible. The natural sciences demonstrate their truths by the fact that they can be applied practically. They establish the foundation upon which all of the technologies and industries that have led to the material wealth and comfort of the Western world are based. The scientific, materialist worldview has become the mythos of our times.

But although this worldview is undisputed, it encompasses only half of reality, only that part which can be measured. All of the invisible dimensions of existence that are not understood by physics and chemistry are missing from the scientific, materialist worldview, including the most meaningful attributes of human beings. Love, joy, beauty, creativity, ethics, and morality cannot be weighed nor measured, and they are therefore nonexistent in the materialistic, scientific worldview.

The indisputable truths and reality that have been established through scientific research are considered universally valid. Awareness of the transcendent and religious nature of these truths gained through meditative observation and illumination could establish the foundation for a new universal spirituality.

The natural sciences and the mystical experience of the world do not contradict one another. On the contrary, they are complementary; they complete one another to comprise the whole truth and reality of our existence.

I would like to offer two further examples in this regard.

Every higher organism—be it a plant, animal, or human—begins its life as a single cell, as a fertilized ovum. The smallest units of life, from which all organisms are built, are the cells. Scientific research has established that not only do plants, animals, and humans have the same cell structure, to a large extent they also have the same chemical composition.

This knowledge is in harmony with the mystic's experience of the unity of all life—of the integral security of the human being within the living Creation. St. Francis of Assisi saw the truth.

A further example is the process known as photosynthesis. With light as the original cosmic source of energy and chlorophyll as the catalyzer, the plant is able to produce an organic substance—our food—from water and carbon dioxide. In the process of human digestion, the food is broken down again into carbon dioxide and water, and the same amount of energy is released and made available to the body as had been taken in during photosynthesis. All things are built and maintained with light as the source of energy. The thought processes of the human brain are also fed from this energy source, so that the human spirit and mind—our consciousness—represent light's highest, most sublime stage of energetic transformation.

We are "light beings." This statement is not just an allusion to the "enlightenment" of mystical experience or to the meaning of light in many religions—it also reflects a scientific insight.

Sunlight is not only the bio-energetic basis for all life on Earth, it is also the medium with which the

Creator makes the wonder of his creation visible to his creations.

The natural sciences also have explained the mechanism of our faculty of vision. They have shown that the image of the colorful world, as we see it, does not exist in the external world; the picture screen is inside us, in our consciousness. Thus, all human beings carry within themselves their own personally created image of the world. Through seeing, through perception, we gain possession of the world. In the existential sense discussed earlier we could gain possession of the entire world. But most of the time our senses—our "gates of perception"—are narrowed and dulled, and thus we lose track of the possessions the Creator had in mind for us.

In the blessed moments, however, we see the whole truth, and we become aware of the entire splendor and nobility of Creation and of our inherent role in its becoming and dying in timeless existence. At such moments we experience what enlightened people have recognized as the purpose of our existence: happiness.

Such a spontaneous, joyful awareness is rare and seems to happen to few people.

But the ability to have a visionary experience must belong to the nature of the human mind, otherwise no one would be able to experience it. The relationship between the happiness of an adult with their childhood experience of the world also explains the innate ability we have for forgiveness.

Children still live in paradise, "for theirs is the kingdom of heaven," says *the* enlightened one.

Children still live in the wholeness of being; the "I" in their consciousness has not yet separated itself from the "you," from the external world—in adults this "I" can grow into the egoism that leads to feelings of separation, loneliness, perdition, and insecurity, with all the unfortunate consequences these have for the fate of the individual.

The dualistic, titanlike, single-minded worldview that sees only what is made by humans and only recognizes this as the true reality is the primary cause for the current worldwide ecological, economical, social, and spiritual crisis. It would be pointless and dumb for anyone to disagree with the reality that everyone experiences daily and hears and sees through mass media.

But, as I've said, this represents only half of reality. What is therefore urgently necessary for shifting the situation today is an unobstructed, clear, and open view, through which we can once again become aware of the totality of the world, of Creation and our integral place within it.

Various methods of meditation have been developed as aids to attain a visionary experience: yoga, fasting, breathing exercises, isolation, and so forth. A particularly effective method is the use of entheogenic drugs to shape meditation, because the pharmacological effects of these psychoactive substances bring about an enormous increase of sensory perceptions, especially in terms of sights and sounds, as well as a change in consciousness in the sense of an expanded and increased sensitivity.

Because the heights and depths of existence are experienced at a level of unfamiliar intensity during a meditation using entheogenic drugs, the danger exists that what one has experienced cannot be integrated into normal consciousness in a meaningful way.

For this reason ancient cultures traditionally use entheogenic drugs in the context of a religious and ceremonial framework. Then the experience can become one that humans have intensely sought since time immemorial: a *unio mystica* with all the happiness that this entails.

Whoever has been blessed once with a moment in which their eyes—both external and internal—were opened has the ability to become aware of the wonder of Creation in their everyday consciousness.

The Greeks of antiquity called the Creation *kosmos,* which means "jewel." Humans had not yet dirtied the world. Today we only experience our world as a jewel from the cosmic point of view—an image provided to us by space research—as the blue planet glowing in the sunlight, floating since time immemorial on its predetermined path in the infinity of the universe.

Those whose eyes are open can see that, despite considerable destruction, primordial life still exists on this wondrous spaceship: the secret world of the oceans, the green continents, and the beauty of its marvelous creations in the plant and animal world.

But most of the time we view the world with dull eyes and with senses that have been numbed through routine, seeing only the part of reality made by human hands, and we search only in this—like searching in a mandala we have made by ourselves—for happiness and meaning.

It would be better to look into the calyx of a flower, of a blossom, which is a thousand times more beautiful in completion and beauty than anything made by humans—for it is filled with life. It is filled with the same life as the person observing it; both the observer and the observed are manifestations of the selfsame spirit of Creation.

As a conclusion I will offer a brief consideration of sight, which plays such a large role in our experience of the world. Two quotes are relevant in this regard. From St. Augustine comes the sentence "our entire reward is vision," and Goethe recognized that we are "born to see, and driven to look."

Seeing develops into observing over three stages that can be differentiated. The first stage consists of the raw perception of an object without it awakening our interest.

The second stage consists of the object drawing our attention to it. In the third stage, the object is more carefully looked at and investigated. This is where thought and scientific research begin.

The highest stage of seeing—of making a connection more generally to an object and to the external world overall—is achieved when the borderline between subject and object, between observer and observed, between myself and the external world, is consciously lifted when I am unified with the world and its spiritual primal source. That is the condition of love.

Life is a hint of the Eternal.

4

The Use of Psychedelics for the Great Transition

A lecture presented at the Second Symposium of the European College for Consciousness Studies, June 12–14, 1987, in Kandern, Germany.

A woman first made the suggestion that psychedelics could be used by people who were dying. In a 1957 essay in the American magazine *This Week* about research on the magic mushroom of Mexico, Dr. Valentina P. Wasson, the wife of the famous ethnomycologist R. Gordon Wasson (who had identified the psychoactive elements of this mushroom), wrote that they now had a medicine available that could not only be used for the treatment of psychological disturbances and alcoholism but also as a source of pain relief for the dying.

The next suggestion regarding the use of a psychedelic for those on their deathbed came from the writer and philosopher Aldous Huxley. In a 1958 letter to the British psychiatrist Humphry Osmond, he wrote about "yet another project—the administration of LSD to terminal cancer cases, in the hope that it would make dying a more spiritual, less strictly physiological process."

In his final book, *Island,* published in 1972, Huxley wrote about a *moksha medicine* (the word *moksha* means "liberation" and "enlightenment" in Sanskrit) that was made from a mushroom. On the island of Pala, where this utopian novel takes place and where an advanced culture had developed from Eastern wisdom and Western civilization. Moksha medicine is used three times in life: for initiation rites at puberty, for midlife crises, and on the deathbed.

When Aldous Huxley lay dying on November 22, 1963, he asked for the moksha medicine that his wife, Laura, administered in the form of 0.1 mg of LSD.

Independently of Mrs. Wasson's suggestions and the case of Aldous Huxley, research on the use of LSD for critically ill patients was conducted in the early 1960s at the Chicago Medical School. There Dr. Eric Kast and colleagues were searching for an effective substance to treat the most difficult cases of pain and came upon the idea to use LSD in their research, because it was known that this substance induced changes in sensual perception. In a public study in 1964, Kast and Collins reported about their findings of LSD compared with Demerol and Dilaudid as pain medication for the most difficult situations. In many cases they determined that the LSD was superior, and they further observed that individual patients experienced a palpable lack of concern about the serious state of their health. In a subsequent study published in 1966, they specifically researched the pain-relieving effects of LSD on cancer patients with only a short life expectancy and who were aware of their diagnosis. Besides the relief of pain, researchers observed a decrease in the patient's fear of death, frequently an experience of "happy oceanic feelings," and a clarified religious position in the face of their imminent death.

Eric Kast should be recognized as a pioneer in the medicinal use of LSD as a pain medicine and as a psychological pharmaceutical for the dying, even if his research can be criticized in retrospect for a certain deficiency in its scientific methodology.

Inspired by Kast's publications, Sidney Cohen, a psychoanalyst and psychotherapist in Los Angeles who was experienced with LSD in his psychiatric practice, began research with LSD on cancer patients. He was able to confirm almost all of Kast's findings and expressed the hope of using LSD as an aid to create a technique to transform the experience of dying.

The most comprehensive and methodically constructed tests about the use of psychedelics were subsequently carried out at the Maryland Psychiatric Research Center in the Grove State Hospital in Maryland. The director and the founder of the project was Walter Pahnke, and the work began in 1967. The team also included Dr. Stanislav Grof, who had already worked with LSD in his native Prague. Also on the team was Albert Kurland, the administrative director of the Research Center, as well as the psychiatrist Charles Savage and the psychologist Sanford Unger. The project was divided into three parts.

1. A thorough psychological examination using a specially developed psychiatric test combined with basic mental preparation of the patient and with the permission of the relatives.
2. The entrance. A high dose of LSD, 200 to 600 micrograms, and later occasionally 90 to 150 milligrams of dipropyltryptamine (DPT), was used in order to provoke a peak psychedelic experience. This included prudent care of the patients and was accompanied by specially chosen music.
3. Psychiatric processing of the experience and evaluation with special tests.

The patients were selected at Sinai Hospital. All were cancer patients with a life expectancy of at least three months, so as to make a scientific evaluation of the psychedelic treatment possible. Usually only a single treatment was employed, and only in a few cases was the treatment repeated. The effects of the LSD and/or the DPT treatment varied greatly from individual to individual, and the nature of the experiences also varied considerably in style and complexity. They included a decrease of depression and fear, an alleviation of pain that sometimes lasted for weeks, but above all, most patients experienced a new attitude toward life and death, a sense of reconciliation regarding dissatisfactions in their life, as well as an awakening of usually unorthodox religious beliefs, combined with a fearless attitude about the great transition that stood before them. In addition to those patients for whom these positive effects were established in varying strengths, there were also those for which the psychedelic treatment was unsuccessful.

Stanislav Grof directed the Grove Project following the tragic death of Walter Pahnke in 1971.* Together with the anthropologist Joan Halifax, who

*[Walter Pahnke died unexpectedly as a result of a scuba-diving accident. —*Trans.*]

had also worked on the project, Grof published the results of this important research in the book *The Human Encounter with Death*. This book contains the original literature about this research project briefly mentioned here.

The research project at the Spring Grove Hospital was recognized for its outstanding scientific methodology. They established that the physical and psychic state of the terminally ill patient facing death is in many cases improved by psychotherapeutic treatment aided by the use of psychedelics. This justifies the hope that the actual transition to the other land, a land from which no traveler returns, can also be eased and spiritualized. But there is no protocol for this and many unanswered questions remain. Everyone dies alone, and no one reports back to our earthly world about whether dying is better under the effects of psychedelics.

Thus the decision to take the moksha medicine remains a gamble. Every individual must make this decision for themselves, depending on their personal worldview, their intuition and desires, and their beliefs.

REFERENCES

Grof, Stanislav, and Joan Halifax. *The Human Encounter with Death*. New York: E. P. Dutton, 1977.

Hofmann, Albert. "50 Jahre LSD." *Jahrbuch des Europäischen Collegiums für Bewusstseinsstudien 1993/1994*. Berlin: VWB, 1994.

Huxley, Aldous. *Moksha*. Los Angeles: J. P. Tarcher, 1977. Reprinted as *Moksha: Aldous Huxley's Classic Writings on Psychedelics and the Visionary Experience*. Rochester, Vt.: Park Street Press, 1999.

———. *Island*. New York: Harper & Row, 1972.

Huxley, Laura. *This Timeless Moment: A Personal View of Aldous Huxley*. Millbrae, Calif.: Celestial Arts, 1975.

ALBERT HOFMANN

The Boundary Walker

A TRIBUTE BY ROGER LIGGENSTORFER

What a joy and a special stroke of good fate it is for a man to be able to live to such an advanced age with all of his senses functioning well. Albert Hofmann was aware of this—and this awareness was present in every moment of his existence. The place where he lived is called "Rittimate" and lies near the village of Burg in Leimental, Switzerland. It is, in his own words, "a garden of paradise," which he had selected himself about thirty-five years earlier when he retired and which he had designed together with his wife, Anita. He often commented that "next to LSD, it was my second greatest discovery."

Many contemporary personalities visited him there and were profoundly touched by him as well as the place. Albert and Anita were attentive hosts: everything was ready at the appointed time; coffee and cake were already on the table. Naturally his "Eau de Vie," the "spirit" of the Rittimate, which he distilled himself, was never absent.

Up until just a few years ago, Albert had harvested the bounty of his garden, fruits such as plums and cherries, himself. When the dozen cherry trees he planted blossomed each year in the spring, he was as delighted as if it were the first time.

It is unfortunate that Albert won't be able to experience this again—he had been really looking forward to it.

His perception—or as he put it, "what a human being believes to be true"—was unique; his clear gaze, his refined hearing, and his particular joy at the wonders of nature, which he saw as if they were small jewels, were unique. Although Albert was very familiar with the path to his "little bench" on the edge of the Rittimatte forest, near the border stone of France—in this sense too he was a boundary walker—every walk to this spot brought new discoveries for him: "Look at this fantastic flower. This blossom! A wonder of nature! Look over there at the beautiful butterfly." Albert was a seeing person, as he described himself. In a poem he declares: "The highest stage of seeing is love. Love is the highest stage of seeing."

For many people wisdom comes with age. This was especially true in Albert's case. Every encounter with him offered the possibility of experiencing the phenomenon of an eternal youthfulness that was always revealed during the time spent with him. He viewed the world through sparkling eyes, with a quick, lively, and humorous mind, sharply observing what he saw. His deep connection with the Creation, with nature, with animals and humans shimmered in every new encounter and made one feel the greatest joy for life when in his presence. Just two weeks before his death he went to the ear doctor—not because it was hard for him to understand conversations, but because he could not hear the singing of the birds in all its nuances anymore.

During our last walk together to the border stone about three weeks before his death, we visited the resting place of his beloved wife, Anita, who had just died shortly before, at Christmas. With his characteristic lightness of being and his clarity in dealing with reality, Albert said, "Beautiful, soon I will be here to enjoy these gorgeous sunsets too." Albert reflected for a brief moment, turning inward but without slipping into sentimental feelings. He had expressed it himself in one of his poems like so: "To affirm life also means to affirm death, for in essence both are indivisibly united."

This text was first published as an obituary in the May 2, 2008, issue of the *Tages Anzeiger* newspaper (Zurich, Switzerland).

ROGER LIGGENSTORFER is a prolific editor and author of books dealing with psychoactive plants and substances as well as governmental drug policy. His recent books include *Albert Hofmann und die Entdeckung des LSD: Auf dem Weg nach Eleusis* (Albert Hofmann and the Discovery of LSD: On the Road to Eleusis) and *Hoffmans Reisen: Innere und äussere Reisen des LSD-Entdeckers* (Hofmann's Trips: Inner and Outer Journeys of the Discoverer of LSD), both coauthored with Mathias Broeckers. He runs the publishing company Nachtschatten Verlag in Solothurn, Switzerland.

St. Albert and the LSD Revelation Revolution

An Afterword by Alex Grey

On January 11, 2006, the Swiss chemist who discovered LSD, Albert Hofmann, Ph.D., turned one hundred years old. To honor his centennial, a three-day LSD symposium was held January 13, 14, and 15 in Basel, Switzerland. Leading scientific, psychiatric, pharmaceutical, legal, artistic, and mystical voices spoke on the various physiological, personal, social, and spiritual impacts of LSD. Hofmann spoke the first and last evening and was showered with praise and applause by more than two thousand attendees (we also sang "Happy Birthday to You"). Hofmann was swarmed with fans wherever he went, and one of the symposium announcers said, "Dr. Hofmann apologizes that he will not be able to sign everyone's book, because he explained, 'I'm no longer 90.'"

The birthday celebration was an elegant gathering of family, friends, and colleagues held at the Museum of Cultures in Basel. My wife, Allyson, and I were invited because of our association with psychedelic culture and participation in a symposium later that week. Distinguished guests at the birthday gathering spoke in German, but even monolinguistic Americans could understand the reverence and enthusiasm shown in speeches praising Albert Hofmann as a scientist and a sage. A reception followed where invited guests mingled and toasted. I was intrigued to learn that none of the members of Hofmann's large family or any of his relatives, except for his wife, had ever tried LSD. The good doctor has always steered away from advocacy yet has come to feel that some kind of divine intervention or destiny did play a role in his discovery.

Hofmann first synthesized the compound LSD in 1938 while researching ergot derivatives as a chemist for Sandoz Pharmaceuticals in Basel. The substance was tested on lab animals with no interesting results, so, like hundreds of similar test compounds, investigation of this drug was abandoned.

Yet in 1943, at the horrific height of WWII and shortly after Fermi made his discovery that led to the atomic bomb, Hofmann had a "peculiar presentiment" to resynthesize LSD. Hofmann said (in 2008 at the Basel LSD Conference) that he heard the voice of LSD calling him and that never before or since had he heard any similar voice. On the first LSD trip, April 19, 1943, Hofmann discovered the intense psychological vortex of psychedelia and experienced an overwhelming fear of dying, feelings of having left his body, and, later, heavenly kaleidoscopic visions. Worried that he had poisoned himself and not wishing to die in his lab, he took a wild bike ride through the streets of Basel to his home, full of perceptual distortions, not knowing whether he would ever return from his madness. Just as early Christians used secret symbols to connect with followers when they were an outlaw religion, LSD users speak of April 19th as "Bicycle Day." The last element I painted on my portrait of Albert Hofmann was a little bike-riding Hofmann (to the right of the molecule on the front cover), and in honor of the good doctor, I was on LSD as I painted it.

Alex Grey and Albert Hofmann in 2006 with the portrait
St. Albert and the LSD Revelation Revolution

In my portrait of Albert Hofmann, the eye of transcendental spirit in the upper left-hand corner of the painting releases spiralic streams of primordial rainbow spheres of potential, one of which becomes a compassionate alchemical angel whose tears drip down to anoint or "create" the LSD molecule that the doctor holds in his hands, and a shadowy demon, here identified with Nazi power, tugs or pushes at it. LSD opens a visionary gateway to the heart, as shown by the spiral of fractally infinitizing eyes resembling the stripey eye-spheres of the molecule, swirling into the center of the chest. On St. Albert's shoulder blade is a portrait of Paracelsus, the Alchemist of Basel, who is credited with founding modern chemistry five hundred years ago, yet his alchemical goal was to discover the Philosopher's Stone. Alchemy was the art and science of the transmutation of the elements, like turning lead into gold and the identification of the soul of the alchemist with the chemical transformations as a metaphor of their journey to enlightenment. Modern chemistry took the psyche and mystery out of the material weighed and measured world, reducing the world to a heap of atoms. LSD brought psyche back front and center to the chemical material world, which is partly why I believe that LSD is the Philosopher's Stone, the discovery of which, also in the town of Basel, is the result of an alchemical process put in motion by the great Paracelsus.

I put a lot of LSD personalities and symbolism in the aura of Albert Hofmann. Some of these people were Hofmann's friends, like Aldous Huxley, R. Gordon Wasson, Maria Sabina, and Richard Evans Schultes. Each of these people had a special connection to psychedelics. Huxley wrote fearlessly about the psychedelic experience in *The Doors of Perception* and *Heaven and Hell,* which also talks about visionary states and works of art. His dying wish was to be injected with 100 mcs of LSD to assist his transition, and this was noted by his wife, Laura. Wasson brought the magic psilocybin mushrooms to the world by attending the Mexican curandera Maria Sabina's sacred mushroom healing ceremony, then writing about it in *Life Magazine.* Hofmann later analyzed the mushroom and distilled the previously unclassified psychedelic psilocybin.

If humanity is around in one thousand years, I believe LSD and psilocybin will still be important tools for consciousness expansion. These and other psychedelic sacraments have brought the healing miracle of a mystical experience to millions of people and allowed visionary artists to paint the transcendental realms from observation. This has given birth to the visionary art movement, a new universal sacred art practiced worldwide. The great uplifting of humanity beyond its self-destruction is the redemptive mission of art. Infinite gratitude is owed to the gentle sage scientist, Albert Hofmann, who launched a consciousness revolution. The promise of psychedelics for human history was best summarized by Hofmann's own words when he was 101 years old.

> *Alienation from nature and the loss of the experience of being part of the living creation is the greatest tragedy of our materialistic era. It is the causative reason for ecological devastation and climate change. Therefore I attribute absolute highest importance to consciousness change. I regard psychedelics as catalyzers for this. They are tools which are guiding our perception toward other deeper areas of our human existence, so that we again become aware of our spiritual essence. Psychedelic experiences in a safe setting can help our consciousness open up to this sensation of connection and of being one with nature. LSD and related substances are not drugs in the usual sense, but are part of the sacred substances, which have been used for thousands of years in ritual settings. The classic psychedelics like LSD, Psilocybin, and Mescaline are characterized by the fact that they are neither toxic nor addictive. It is my great concern to separate psychedelics from the ongoing debates about drugs, and to highlight the potential inherent to these substances for self-awareness, as an adjunct in therapy, and for fundamental research into the human mind. It is my wish that a modern Eleusis will emerge, in*

which seeking humans can learn to have transcendent experiences with sacred substances in a safe setting. I am convinced that these soul-opening, mind-revealing substances will find their appropriate place in our society and our culture.

—Albert Hofmann
April 19, 2007

Alex Grey is the author of *Sacred Mirrors, Transfigurations, The Mission of Art, Art Psalms,* and *Net of Being,* from which a portion of this text originally appeared. His work has graced numerous album covers including those of Nirvana, TOOL, and the Beastie Boys; appeared in *Newsweek;* and been exhibited throughout the world, including the New Museum in New York City and the Museum of Contemporary Art in San Diego. In 2013, *Watkins Review* placed Alex at #12 on the 100 most spiritually influential living people. In 2004, Alex and his wife, artist Allyson Grey, opened the Chapel of Sacred Mirrors (CoSM), a sanctuary for visionary art and culture, in New York City. In 2009, CoSM, now a church, moved from Manhattan to Wappinger, New York, in the Hudson Valley.

Completion Not Exclusion

What is truth—is it the picture of reality determined for us by the natural sciences, or the one experienced by the mystic in visions? The question can only be asked if we assume that the scientific and mystical experience of the world mutually exclude one another. And this opinion dominates. But this is not the case. On the contrary, the natural sciences and mystical experience complete one another.

From "Naturwissenschaft und mystische Welterfahrung" (Natural Science and Mystical Experience), *Jahrbuch für Ethnomedizin und Bewusstseinsforschung 1992* (Berlin: VWB, 1993), 9.

About the Author

Albert Hofmann was born January 11, 1906, in Baden, Switzerland. After a commercial apprenticeship and passing his high school exit examination, he studied chemistry at the University of Zurich. He graduated with distinctions, and in 1929, when he was twenty-three, he began working in the pharmaceutical and chemical research laboratories at the Sandoz company in Basel, Switzerland. He first worked on elucidating the chemical structure of squill (*Scilla maritime*). Beginning in 1935 he turned his attention to ergot of rye (*Claviceps purpurea*). In 1938 he isolated the basic building block of all therapeutically significant ergot alkaloids, lysergic acid, and researched the effects of the lysergic-acid derivatives. This research resulted in a number of therapeutically valuable medicines that include Dihydergot, which is used for circulation and stabilizing blood pressure; the uterine contracting and styptic Methergine; and the geriatric medicine Hydergine, which improves circulation.

In April of 1943, Albert Hofmann discovered the hallucinogenic effects of LSD. His discovery made him famous among international experts. After this he continued to make significant contributions to LSD research throughout the world. Thus, in 1958, he was the first person to isolate psilocybin and psilocin, the psychoactive ingredients of the Mexican magic mushroom (*Psilocybe mexicana*). His research has appeared in numerous scientific publications and books, including *Plants of the Gods, Insight—Outlook,* and his autobiography, *LSD: My Problem Child.*

Hofmann remained active at Sandoz after his retirement, concluding his time there as Director of the Research Division for Natural Medicines.

After this he dedicated himself to writing and giving lectures. He received honorary doctorates from the Eidgenössische Technische Hochschule of Zurich, the University of Stockholm, and the Freie Universität Berlin. He also served as a member of the Nobel Peace Prize committee. From January 13 to 15, 2006, on the occasion of his hundredth birthday, there was a symposium held in Basel, "LSD—Problem Child and Wonder Drug." Albert Hofmann died of a heart attack on April 29, 2008, in Burg in Leimental, Switzerland, at the age of 102.

BOOKS BY ALBERT HOFMANN

Die Mutterkornalkaloide. Stuttgart: Ferdinand Enke Verlag, 1964. Reprinted Solothurn, Switzerland: Nachtschatten Verlag, 2000.

The Botany and Chemistry of Hallucinogens (with Richard Evans Schultes). Springfield, Ill.: Thomas, 1973. Reprinted 1999.

The Road to Eleusis: Unveiling the Secret of the Mysteries (with R. Gordon Wasson and Carl A. P. Ruck). New York: Harcourt, Brace, Jovanovich, 1978. Reprinted Berkeley, Calif.: North Atlantic Books, 2008.

Plants of the Gods: Origins of Hallucinogenic Use (with Richard Evans Schultes). New York: McGraw-Hill, 1979. Revised and expanded edition (with Christian Rätsch). Rochester, Vt.: Healing Arts Press, 2001.

LSD: My Problem Child—Reflections on Sacred Drugs, Mysticism, and Science. Translated by Jonathan Ott. New York: McGraw-Hill, 1980. [English edition of *LSD: Mein Sorgenkind*] Reprinted Sarasota, Fla.: Multidisciplinary Association for Psychedelic Studies (MAPS), 2009.

Insight—Outlook. Atlanta, Ga.: Humanics New Age, 1989. [English edition of *Einsichten—Ausblicke*]

Lob des Schauens. Burg, Switzerland: Privately printed, 1996. Solothurn, Switzerland: Nachtschatten Verlag, 2002.

Isn't it wonderful,
that we know not,
where we came from,
where we are going?
Knowing would destroy
the wonder.

INDEX

BOOKS OF RELATED INTEREST

Plants of the Gods
Their Sacred, Healing, and Hallucinogenic Powers
by Richard Evans Schultes, Albert Hofmann, and Christian Rätsch

Encyclopedia of Psychoactive Plants
Ethnopharmacology and Its Applications
by Christian Rätsch
Foreword by Albert Hofmann

DMT: The Spirit Molecule
A Doctor's Revolutionary Research into the Biology of Near-Death and Mystical Experiences
by Rick Strassman, M.D.

The Psychedelic Explorer's Guide
Safe, Therapeutic, and Sacred Journeys
by James Fadiman, Ph.D.

The Pot Book
A Complete Guide to Cannabis
Edited by Julie Holland, M.D.

Spiritual Growth with Entheogens
Psychoactive Sacramentals and Human Transformation
Edited by Thomas B. Roberts, Ph.D.

The Psychedelic Future of the Mind
How Entheogens Are Enhancing Cognition, Boosting Intelligence, and Raising Values
by Thomas B. Roberts, Ph.D.

The New Science of Psychedelics
At the Nexus of Culture, Consciousness, and Spirituality
by David Jay Brown